王大伟 著

雨青工作室 绘

[加]Jennifer May等 译

外出隐患

中国水利水电出版社

www.waterpub.com.cn

· 北京 ·

喇叭花

喇叭花，　吹喇叭，

我家有个胖小丫。

胖小丫，　不闻花，

花籽进鼻麻烦大。

Morning Glory

Morning glory's, like a bell.
Look, but try not to smell;
As seeds get into my nose,
I may not feel very well.

**Try not to smell the flowers.
The seed of some flowers might be inhaled into
a child's nose and they might give children allergies.**

小 花 狗

小 花 狗， 小 花 猫，

不 去 逗， 远 远 瞧。

抓 伤 咬 伤 都 没 有，

妈 妈 奖 励 大 香 蕉。

Puppy

A stray dog and cat, two of these.
Don't touch them or tease;
As a scratch could mean rabies,
So hands-off them, please.

Rabies is a very dangerous disease.
Its mortality rate is 100%.
So don't tease wild dogs and wild cats.

汽车安全

汽车不是小宫殿，

孩子不能忘里边。

夏天中暑冬天寒，

遇到窃贼更危险！

小案例
王大伟提示

The Family Car

Left in the car, they forgot.
Winter cold and Summer hot,
What if a thief comes?
Secure and safe I am not.

Don't leave your child alone in the car.

过年鞭炮安全

不在家里放鞭炮，

燃放区边请绕道。

爸爸妈妈看着放，

点了不响勿近瞧。

小案例
王大伟提示

Firecrackers are Dangerous

As the firecrackers appear.
I watch and don't go near.
As adults set them off;
From afar there is no fear.

***When setting off fireworks, parents should
accompany their children to ensure safety.***

健 身 器 小 滑 梯

健 身 器， 小 滑 梯，

生 锈 晃 动 有 危 机。

爸 妈 先 要 查 一 查，

安 全 游 戏 笑 嘻 嘻。

小案例
王大伟提示

Fitness Equipment

Fitness equipment must survey,
Check if it's secure before play,
Parents should check it carefully;
Safety first and then I can enjoy.

Always check the safety of fitness or entertainment equipment before letting the children play.

12·

婴儿车安全

婴儿车， 有风险，

河边扶梯不安全。

爷爷奶奶推车去，

嘱咐不要站河边。

The Baby Carriage

Don't take your baby carriage to an unsafe place.
Be careful near the river or on the escalator, a slow pace,
When grandparents go out, they need to know;
If they are near the river or escalator, leave space.

Parents should keep the baby carriage away from the riverbank, escalator, and steps.

电 梯 安 全 歌

扶 梯 张 开 老 虎 口，

咬 坏 脚 丫 咬 断 手。

别 把 电 梯 当 玩 具，

人 多 等 会 咱 再 走。

Escalator Safety

The escalator is not to enjoy.
No playing, it's not a toy,
I don't want to get hurt;
Listen, every girl and boy.

***The rolling escalator is not a playground or a big toy.
It has many potential dangers and is likely to hurt children.***

小案例
王大伟提示

安 全 倒 车

倒 车 也 是 大 灰 狼，

宝 宝 不 要 站 一 旁。

车 库 车 场 不 玩 耍，

小 孩 小 狗 要 避 让。

小案例
王大伟提示

The Parking Garage

The big bad wolf and the car are alike.
Reversing cars could strike;
No playing in the parking lot or garage.
Avoid the puppies and the little tyke.

Children should not play in the parking lot.

突 发 事 件

不 捑 发 光 小 宝 石，

人 多 拥 挤 脚 步 止。

楼 梯 桥 洞 易 踩 踏，

假 山 慎 防 落 下 石。

小案例
王大伟提示

Radiation Emergency

Unsafe stones and gems that glow,
I can't pick them up, I know!
Falling rocks can hurt me;
And the crowds just don't go.

Don't pick up small shiny gems,
because they're probably radioactive.
And pay attention to the rockery in the park.
The stone on it might fall and hurt you.

旋 转 门

旋 转 门， 会 卡 手，

手 扶 梯， 咬 指 头。

伤 过 很 多 小 朋 友，

妈 妈 领 我 躲 着 走。

小案例
王大伟提示

The Revolving Door

Don't play with the revolving door, it's a no.
The escalator might bite your finger, I know.
They have hurt the hands and fingers of many kids.
Mothers stop children from playing and just go.

Don't let the children play in the revolving door.

小宝宝少飞行

小宝宝，还很小，

坐飞机，爱哭闹。

小手不断把耳挠，

心中一定很烦躁。

小案例
王大伟提示

Less Flying for the Babies

Babies can't fly they are too small.
Flying means crying for the baby doll.
The poor baby's ears will hurt.
The baby will be upset, so no flying at all.

Children should fly less because it could hurt their ears.

小 气 球

小 气 球， 升 上 天，

小 朋 友， 手 别 牵。

站 在 一 边 笑 着 看，

爆 炸 着 火 惹 麻 烦。

小案例
王大伟提示

Small Balloons

Small balloons up in the sky.
I should just watch them fly,
It could pop or catch fire;
It's a new rule to apply.

The hydrogen balloon contained flammable material.
It might cause serious harm to the children.

上学篇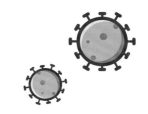

一、交通小病毒

1. 过马路不理睬红灯还通行

2. 过马路时不拉着大人的手，行走时不专心

3. 在马路上滑旱冰（没有刹车，危险）

4. 独自在道路上玩耍、坐卧

5. 过马路不走斑马线，钻越、跨越交通护栏或道路隔离设施

6. 坐车时，把头、手伸出窗外

二、校园小病毒

1. 学校周围有各种小摊和小商店，人员流动性大
2. 天黑时不与陌生人保持距离，平时与陌生人距离太近
3. 家长只送孩子到学校门口，不等孩子安全进校门就转身离开
4. 上学、放学时，学校门口人多，不避开门口拥挤的人群
5. 放学不与小朋友结伴而行
6. 上学、放学时与陌生人说话，吃陌生人的东西

三、性侵害小病毒

1. 孩子，尤其是小女孩经常一人在家
2. 爱接受陌生人给的东西
3. 放学被陌生人接走
4. 家长把孩子交给半熟脸看管
5. 家长没有察觉异性老师的怪异行为（例如对学生又摸又亲）
6. 家长没有警惕邻居大叔对孩子的过分举动（教孩子小裤衩不许别人摸）

图书在版编目（CIP）数据

王大伟儿童安全童谣. 外出隐患 : 汉英对照 / 王大
伟著. -- 北京 : 中国水利水电出版社, 2021.9
　　ISBN 978-7-5170-9617-7

Ⅰ. ①王… Ⅱ. ①王… Ⅲ. ①安全教育－儿童读物－
汉、英 Ⅳ. ①X956-49

中国版本图书馆CIP数据核字(2021)第092370号

责任编辑　　李格（1749558189@qq.com　010-68545865）

书　　名	王大伟儿童安全童谣：外出隐患
	WANG DAWEI ERTONG ANQUAN TONGYAO：WAICHU YINHUAN
作　　者	王大伟　著
绘　　图	雨青工作室
英文翻译	[加]Jennifer May　王大伟　陈诗楠　刘原
配音朗读	王许瞳　李晟元　郑方允　崔璎峤　侯清芸　吴郁暖　钟璇　郑淑予
出版发行	中国水利水电出版社
	（北京市海淀区玉渊潭南路1号D座　100038）
	网址：www.waterpub.com.cn
	E-mail：sales@mwr.gov.cn
	电话：（010）68367658（营销中心）
经　　售	北京科水图书销售中心（零售）
	电话：（010）88383994、63202643、68545874
	全国各地新华书店和相关出版物销售网点
排　　版	韩雪
印　　刷	天津久佳雅创印刷有限公司
规　　格	210mm×190mm　24开本　5印张（总）　120千字（总）
版　　次	2021年9月第1版　2021年9月第1次印刷
总 定 价	68.00元（全4册）